DANGEROUS JOURNEYS

EXPLORING ANTARCTICA

BY ALLAN MOREY

TORQUE™

BELLWETHER MEDIA · MINNEAPOLIS, MN

Torque brims with excitement perfect for thrill-seekers of all kinds. Discover daring survival skills, explore uncharted worlds, and marvel at mighty engines and extreme sports. In *Torque* books, anything can happen. Are you ready?

This edition first published in 2023 by Bellwether Media, Inc.

Library of Congress Cataloging-in-Publication Data

LC record for Exploring Antarctica available at: https://lccn.loc.gov/2022012978

Editor: Kieran Downs Designer: Josh Brink

Printed in the United States of America, North Mankato, MN.

TABLE OF CONTENTS

THE SOUTH POLE

Ice crunches under the explorers' boots. A frozen landscape lies in front of them. They are journeying across Antarctica, the coldest place on Earth!

SOUTH POLE

The explorers have been hiking for hours. They are hungry and **exhausted**. Yet, they push on. They are attempting to reach the **South Pole**.

THE COLDEST PLACE ON EARTH

The **continent** of Antarctica lies at the bottom of the world. It covers more than 5 million square miles (13 million square kilometers).

Antarctica has a harsh **climate**. Temperatures often drop below freezing. Winds can gust up to 200 miles (322 kilometers) per hour.

ICY DESERT

ANTARCTICA IS THE WORLD'S LARGEST DESERT. IT RECEIVES VERY LITTLE RAIN OR SNOW EACH YEAR.

ANTARCTICA MAP

SOUTH POLE = ◆

ANTARCTICA = ◆

In 1821, John Davis was likely the first person to set foot on Antarctica. But exploration of the continent did not begin until years later.

In the early 1900s, several explorers tried to reach the South Pole. Some even died in their attempts. Then in 1911, Norwegian Roald Amundsen set foot on Earth's southernmost point.

ROALD AMUNDSEN

ROALD AMUNDSEN AT THE SOUTH POLE

NOTABLE EXPLORER

NAME: ANN BANCROFT

BORN: SEPTEMBER 29, 1955

JOURNEY: SET OFF ON THE AMERICAN WOMEN'S EXPEDITION TO ANTARCTICA IN 1992

RESULTS: ALONG WITH THREE OTHER EXPLORERS, SKIED 660 MILES (1,062 KILOMETERS) TO BECOME ONE OF THE FIRST WOMEN TO REACH THE SOUTH POLE

Today, people mostly travel to Antarctica for science. Some study its wildlife. Penguins, seals, and orcas live along the coast. Others study the effects of **climate change**. They learn how Antarctica is changing.

KAYAKING

Other people travel to Antarctica as **tourists**. They hike its landscape. They kayak in its waters.

PLANNING AND PREPARATION

Antarctica is not an easy place to reach. It is more than 600 miles (966 kilometers) from the nearest city. There are no regularly scheduled flights to the continent. Most people arrive by boat.

BOTTOM OF THE WORLD

USHUAIA, ARGENTINA, IS THE CLOSEST CITY TO ANTARCTICA. IT IS FOUND AT THE SOUTHERNMOST TIP OF SOUTH AMERICA.

Some governments send scientists to Antarctica. Most tourists must arrange travel through a cruise operator.

Antarctica does not belong to any one country. It is managed by the **Antarctic Treaty System**. The United States and 11 other countries created this agreement.

Travelers must get permission from their government to visit Antarctica. They need to agree not to disturb wildlife. They cannot cause damage to the **environment**.

MEETING ABOUT THE ANTARCTIC TREATY SYSTEM

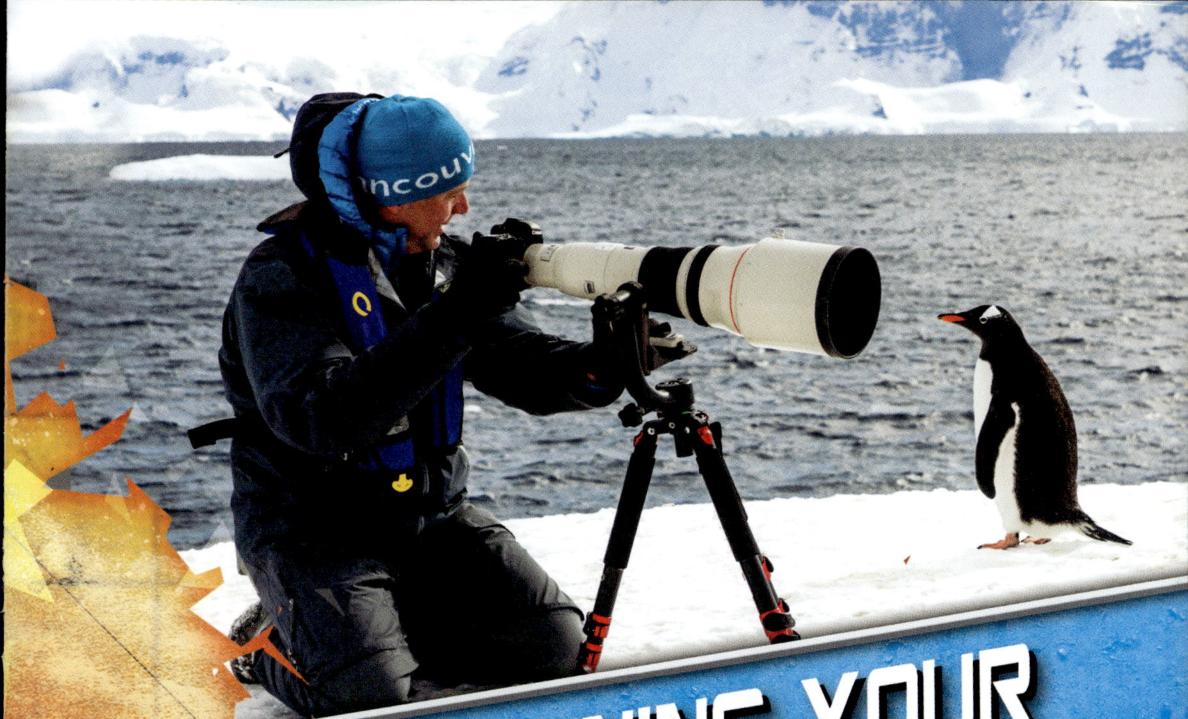

PLANNING YOUR JOURNEY

ON THE JOURNEY

People traveling to Antarctica must prepare for its **extreme** climate. **Frostbite** and **hypothermia** can develop quickly. Waterproof clothing helps keep people warm.

Sunlight **reflects** off snow and ice. It can cause snow blindness and sunburns. Explorers wear goggles or sunglasses to protect their eyes. Sunscreen protects their skin.

WHAT HAPPENS WHEN YOU HAVE HYPOTHERMIA?

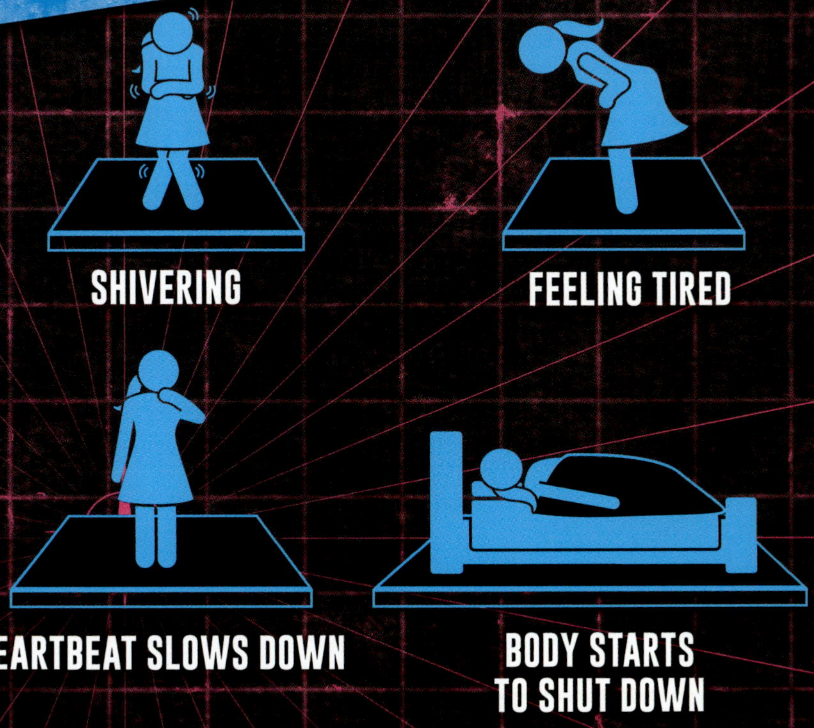

There are no roads to travel around Antarctica. Scientists often get around by helicopter or snowmobile. They also often travel on cross-country skis. They make maps using **GPS**.

MIGHTY MOUNTAIN

MOUNT VINSON IS THE TALLEST MOUNTAIN IN ANTARCTICA. IT STANDS 16,050 FEET (4,892 METERS) TALL.

But deep **crevasses** line Antarctica. People and snowmobiles can easily fall into them. Scientists use **radar** to help find where the cracks are.

Exploring Antarctica can be dangerous. But brave people who are prepared have a chance to see some incredible sights.

Mountainous **glaciers** line the continent's coasts. Rare wildlife swims off its shores. Towering mountains fill the horizon. A journey to Antarctica rewards explorers with unforgettable memories!

GLOSSARY

Antarctic Treaty System—a series of agreements between countries that set the rules for how Antarctica is managed

climate—the average weather of a place over a long period of time

climate change—a human-caused change in Earth's weather due to warming temperatures

continent—one of Earth's seven large land masses

crevasses—deep cracks in the ground or ice

environment—the land, air, plants, water, and animals in an area

exhausted—physically weak or tired

extreme—beyond what is usual or might be expected

frostbite—damage to the body from being frozen for a long period of time

glaciers—large, slowly moving masses of ice

GPS—global positioning system; GPS is a system of satellites and devices that people use to find locations.

hypothermia—a condition in which the body loses heat faster than it can produce it; hypothermia causes body systems to shut down.

radar—a system that tracks the location of an object

reflects—bends or throws back waves of light

South Pole—the southernmost point on Earth

tourists—people who travel to visit other places

TO LEARN MORE

AT THE LIBRARY

Green, Sara. *Deserts.* Minneapolis, Minn.: Bellwether Media, 2022.

Hoena, Blake. *Trapped in Antarctica: The Story of Shackleton and the Endurance.* Minneapolis, Minn.: Bellwether Media, 2022.

Kellaher, Karen. *Antarctica.* New York, N.Y.: Children's Press, 2019.

ON THE WEB

FACTSURFER

Factsurfer.com gives you a safe, fun way to find more information.

1. Go to www.factsurfer.com
2. Enter "exploring Antarctica" into the search box and click 🔍.
3. Select your book cover to see a list of related content.

INDEX

The images in this book are reproduced through the courtesy of: Travel_Photoshoot, front cover (hero); Volodymyr Goinyk, front cover (background), pp. 3, 6; Yongyut Kumsri, front cover (small explorer), p. 3; I. Noyan Yilmaz, front cover (tent), pp. 3, 23; tandemich, front cover (ice), p. 3; Images&Stories/ Alamy, p. 4; ZUMA Press/ Alamy, pp. 5, 9 (Ann Bancroft); blickwinkel/ Poelking/ Alamy, p. 7; Everett Collection/ Alamy, p. 8; North Wind Picture Archives/ Alamy, p. 9; Paul Glendell/ Alamy, p. 10; Have Camera Will Travel | Antarctica/ Alamy, p. 11; Sergio Pitamitz/ Alamy, p. 12; evenfh, p. 13; Frank Barratt/ Stringer/ Getty Images, p. 14; H. Mark Weidman Photography/ Alamy, p. 15; reisegraf.ch, p. 15 (boat journey, wildlife); Artman96, p. 15 (permission); Lucky Business, p. 15 (pack gear); World Travel Collection/ Alamy, p. 16; Leremy, p. 17 (icons); Anadolu/ Getty Images, p. 18; Holger Leue/ Getty Images, p. 19; Song_about_summer, p. 20; Philipp Salveter, p. 21.